Experimental Design Process

A Guide for High School Students to Conduct STEM Research Like a Research Scientist

Why I Wrote the Experimental Design Book?

Judging the Intel International Science Fair in May 2018

As a girl growing up in Miami, I always loved Math and Science. The adults in my life encouraged my love of Math and Science. It was in 10th Grade, in Ms. Shy's Chemistry class that I discovered that chemistry was my favorite STEM discipline. The main reason that I loved chemistry was because of the experiments. While completing my Chemical Engineering Degree at Florida A & M University (FAMU), I conducted research with a Chemistry Professor for 3 years. During the 3 years that I conducted research, I learned that research was much more than conducting your experiment. In these 3 years, in addition to conducting experiments, I developed scientific literacy, laboratory and research skills and oral and written communication skills. The skills that I developed while conducting researcher were transferable to other aspects of my career.

In 2005, when I started my nonprofit Science, Engineering and Mathematics Link Inc. (SEM Link), one of our first programs was to develop young STEM Researchers. From Day 1, our programing focused on preparing students for STEM Fairs; that early programming developing into SEM Link's Experimental Design Program. Our Experimental Design program supports STEM Fairs by providing members of the STEM Community to serve as judges for STEM fair; which I make time in my schedule to serve as a judge. This program also supports young STEM Researchers through group mentoring that increases the quality and quantity of STEM fairs. This program's curriculum provides opportunities for K-12 students with strategies to develop their scientific literacy, research and laboratory skills as well as make their projects more competitive at their school, district, regional and higher-level fairs and competitions. This curriculum was developed and implemented initially at school sites working with student at our school partners. We also had a two-page Experimental Design Handbook available on our website to support students.

In order to expand the impact of this curriculum, two years ago I received funding from the Jack and Jill of America, to turn the 2-page Experimental Design Handbook into an online course. This year, I was asked to make this content in the online course into a book. Therefore, I honored this request to make the content in the online course available in this book.

My love of supporting young STEM Researchers is the reason I developed the Experimental Design Process curriculum; the content for the online course and this book. I am believer that STEM Research is where youth get an opportunity to explore their interest in a STEM discipline/topic. STEM is all about being curious, asking questions and doing research, conducting experiments and collecting data to answer their questions. In the K-12 educational setting, STEM Fairs are the venues for kids to do explore that curiosity, so I am supporter of youth participating in STEM Fairs. However, kids aren't adequately prepared to conduct STEM Research and make their STEM fair projects competitive. There are a lot of reasons for this, but the main reason is that as great as they are educators aren't STEM Researchers neither are most parents.

As a STEM professional, that has conducted research, I feel like it our responsibility as the STEM Community to support the K-12 students that want to conduct STEM Research. As STEM professionals during our education and careers, we developed skills that educators didn't that allow us to help youth conduct STEM Research and prepare for STEM fairs. The content in this book, is the knowledge that I learned as a STEM Researcher, that I adapted for high school students. The reason that I focus on high school students is because of their ability to enter competitions beyond their school and school district fairs. High school students can compete in high-level STEM competitions such as the Intel International Science Fair. These higher-level STEM competitions provide high school students with an opportunity to win cash prizes for their research. It also provides opportunities for them to explore post-graduation opportunities. My goal as a STEM Educator ensure that students are equipped with the tools to have a STEM Fair that is can compete for the international science fairs, whether they conducted their experiment at home, in a lab or in the field.

Table of Contents

Introduction to Experimental Design………………………………… Page 5

Selecting Your Topic…………………………………………….. Page 8

Conducting Background and Scholarly Research……………………. Page 14

Your Hypothesis……………………………………………Page 18

Designing Your Experiment…………………………………. Page 22

Conducting Your Experiment……………………………………Page 30

Writing Your Report………………………………………Page 37

Oral Presentation ……………………………………………Page 42

INTRODUCTION TO EXPERIMENTAL DESIGN

Module 1
Experimental Design Process

Introduction to Experimental Design

The goal of this book, which was adapted from SEM Link's Experimental Design Process online course, is to teach high school students the skills that are needed to prepare for a Science, Technology, Engineering and Mathematics (STEM) Fair, as a research scientist would prepare for presenting their research at a conference. You can read and complete the exercises in this book at your own pace; however, because the knowledge builds as your complete each module it is suggested that you complete each module in sequential order.

Here are a few strategies to allow you to get the most out of reading this this book in order to help you prepare for your upcoming STEM fair.

1. Complete all the exercises in each module and answer the questions
2. If there is a concept that you don't understand while reading this book and completing the exercises in the book, be sure to conduct additional research to help you learn about and understand the concept.
3. There are exercises in this book that as you complete them you will be completing part of your written report for the STEM fair. Be sure to complete those exercises in Google Docs or Word so that you can simultaneously work on your written report reading this book.

What is Experimental Design?

Experimental Design is the process of planning your experiment prior to conducting your experiment. It includes:

- Learning the scientific theory connected to your experiment
- Selecting the suitable environmental conditions to conduct your experiment
- Determining what materials and equipment are needed to conduct your experiment
- Determining your control, dependent and independent variables
- Writing your step by step procedure and safety guidelines for your experiment.

Why Use Experimental Design?

The Experimental Design Process allows you to do more than just conduct an experiment to enter in a STEM fair; but it will help you become a STEM Researcher. The benefits of using the Experimental Design Process as you prepare for your STEM Fair are:

- You will develop your STEM Literacy Skills by learning how to conduct scholarly research, read journals and other scientific publications.
- You will learn the scientific theory connected to your project. A STEM project is more than just conducting an experiment and putting it on a project board. Every experiment is connected to scientific theory and principles. Understanding and explaining the scientific theory and principles will help you write your hypothesis, predict and explain your experimental results.
- You will learn to plan your experiment prior to conducting your experiment. When you plan your experiment prior to conducting the experiment, you will get better experimental results. Planning will allow you to gather everything you need to conduct your experiment and know how you will conduct your experiment.
- You will learn how to design and conduct an experiment that is reproducible; which is a requirement in STEM research.
- You will learn strategies for preparing a competitive oral presentation and written report for the STEM fair that will increase your chances of getting good scores from judges and qualifying to compete at the next level fair.

End of Module Quiz

1. **What is Experimental Design?**
   - It is the steps you take when planning your experiment
   - It is the process of reflection once your experiment is complete
   - It is how you determine what materials and equipment are needed to conduct your experiment

2. **Why should you plan your experiment ahead of time?**
   - It allows you to prepare a great oral and written presentation at STEM fairs
   - It will allow you to collect better data and have an experiment that is reproducible
   - All of the above

SELECTING YOUR TOPIC

Module 2
Experimental Design Process

Selecting Your Topic

The first step in completing your project for the STEM fair is selecting your topic. While preparing for a STEM fair you will spend time working on a project in a topic in which you are interested in investigating further by conducting an experiment.

In the space below, if you need more space write down in Google Docs, Microsoft Word or a Notebook, your answers to the following questions that you guide you through the process of selecting a topic for the fair.

1. Think about one of your favorite hands on STEM activities that you conducted at school, in an after-school program, summer program and/or community event? Describe the activity. What made it fun? What did you learn from the activity?

2. What is your favorite STEM class? Why is that your favorite STEM class?

3. Do you watch any TV shows that are STEM based and/or feature STEM topics? What is the name of the show(s)? What is the show about? Why do you like the show?

4. Name a few of your favorite books and/or YouTube Channels? Are any of them STEM based? Why do you like them?

When you answer the questions in Exercise 1, Did you find anything in common with the STEM activity, your favorite STEM class, the TV shows books and/or TV Channels. For example, did all your activities have animals in them? Were they all about forensic science? Did they have robots? In the space below write down the things that you found in common:

If you have found similarities, that is a topic that you are interested in and should select a project in this topic. In the space below, write down the topic that you are interested in studying for your STEM fair project.

Selecting Your Project

Now that you know what STEM topic you are interest in, you can select your topic and project for your STEM fair project. Here are ideas for selecting a project for your STEM fair:

- Create Your Own: Based on a project that you've always wanted to do, a question that you want to answer or an experiment that you want to conduct
- Research STEM Fair Projects Online
- Find a Project in a STEM Fair Project Book

Exercise 1: Writing Down Your Project Idea

After you have utilized one of the 3 suggestions to find your STEM fair project idea, in the space below (if you need more space write in down in Google Docs, Microsoft Word or notebook), write down your project idea be sure to include the following things:

- STEM Discipline of Your Project:

- Project Title

- Project Question- The question you want to answer with your experiment

Exercise 2: Project Description

In the space below, if you need more space write in down in Google Docs, Microsoft Word or notebook, write a brief description of why you want to and how you will answer your question with your experiment.

End of the Module Quiz

1. **What is one way you can select your science fair project, according to the slides in this module?**
   - Research STEM Fair projects online
   - Find a project that was successfully completed by someone at your school in previous years and copy their information
   - Find a project that is easy to do

CONDUCTING BACKGROUND RESEARCH AND SCHOLARLY RESEARCH

Module 3
Experimental Design Process

Conducting Background Research and Scholarly Research

Why Conduct Background Research?

Now that you have selected your STEM fair topic and project it is time to start conducting background research. Conducting Background Research is an important, but often overlooked part, of the STEM Fair project. In the space below write down your thoughts on why it is important to conduct background research for your STEM Fair Project.

Reasons Why You Must Conduct Background Research for your STEM Fair Project:

- Your STEM Fair project is more than the experiment that you conduct; every experiment is connected to a STEM discipline that has STEM theory. It is important that you understand and can explain the scientific theory connected to your experiment.
- Conducting background research helps you design your experiment and predict experimental results.
- Conducting background research helps you write several components of your research paper for your experiment and parts of your written report.

What is Scholarly Research?

Scholarly Research is research that is written by experts in a field. It contains the following components:

- Written by a scholar or expert in the field that has credentials and affiliations
- All sources in the article are cited
- The articles are peer-reviewed
- Presents research findings or expands knowledge in a specific discipline or field of study
- Has discipline specific jargon and/or technical terms
- Has tables and/or graphs that display research data

Finding Scholarly Research?

Where do you find the scholarly research?

- Your local public, school or college/university library
- Scientific journals and magazine
- Talking to experts in the field
- Internet Research at Scholarly sources/websites

Conducting Background Research

When conducting your background research, make sure that you find the following things to get what you need for your background research:

- Theories connected to your experiment
- Data and charts that can explain the theory and can be included in your research report
- Any articles that can help you form your hypothesis and predict your experimental results
- Technical terms that are connected to your topic

Exercise 3: Start Conducting Your Background Research

As you are conducting your background research, make sure that you get a notebook and/or Google Docs to record the following things to document your research:

- Technical terms/Vocabulary for your topics
- Take notes from the articles and/or books that you can read to learn more about your topic
- Be sure to save the articles and/or pages in the books (make copies of the pages in the book if you check it out from the library) that you used to learn your theory.
- Be sure to highlight any part of the article and/or books that you think you may put in your research paper when you begin writing it.

End of Module Quiz

1. Why is it important to conduct background research?

- The process will help you write parts of your research paper for the experiment
- It helps you understand and explain the scientific theory connected to your experiment
- All of the above

2. Which of the following components are NOT considered to be a characteristic of scholarly research?

- Peer reviewed articles
- Popular internet articles
- Written by a credentialed expert in the field

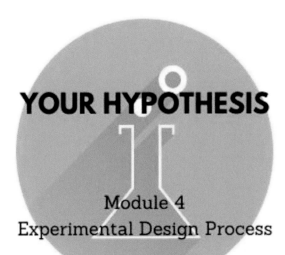

YOUR HYPOTHESIS

Module 4
Experimental Design Process

Your Hypothesis

What is a Hypothesis?

A hypothesis is a tentative and testable answer to a scientific question. In your hypothesis statement, you make a prediction about what you think will happen in your experiment based on what you have read and studied from your background research on the topic of your research/scientific question/STEM Fair project.

What Makes a Good Hypothesis?

- It is based on information from scholarly research about the topic
- At least one clear prediction about the experimental results can be made from the hypothesis
- The predictions that result from the hypothesis must be testable
- The predictions should have both an independent variable (something you can change) and a dependent variable (something you can observe or measure)

Why Write a Hypothesis?

The hypothesis provides a suggested solution for your experiment based on your background research about your topic. Writing your hypothesis does two things:

1. Show that you are familiar with the scientific theory behind your experiment
2. Allows you to keep your experiment scientific by utilizing scientific theory to guide your experimentation

How Do Your Form A Hypothesis?

1. Generate a simple hypothesis by writing down your initial idea about your variables and how they might be related according to scientific theory in a simple declarative statement
2. Make predications by writing down at least one prediction which results from your hypothesis

Exercise 4: Experimental Questions

At this point, you should have conducted your background research which makes you more knowledgeable about the scientific concepts and theories connected to your research/STEM fair topic. Use your literature (articles, webpages, etc.) to write a list of 2-5 questions in your Google Docs/notebook, that you want answer about your topic that you think that collecting data from your experiment will answer.

Question 1:

Question 2:

Question 3:

Example of a Hypothesis & Prediction

Hypothesis:

Electric Motors work because they have electromagnets them, which pull/push on permanent magnets and make the motor spin. As more current flows through the motor's electromagnet, the strength of the magnetic field increases, thus turning the motor faster

Prediction:

If I increase the current supplied to an electric motor, then the RPM's (revolutions per minute) of the motor will increase

Exercise 5: Determining Your Independent and Dependent Variable

Determining Your Independent and Dependent Variables are important for you to write your hypothesis and design your experiment. In the space below, if you need more space write in your notebook/Google docs, write down your

Independent Variable:

Dependent Variable:

Your Hypothesis and Your Experiment

When writing your hypothesis don't worry about if it will be right or wrong. The point of your experiment is not to prove if your hypothesis is right. You write your hypothesis based on the knowledge of the topic of your experiment from your background research. While conducting your experiment, you will collect data that will either support or not support your hypothesis.

Exercise 6: Write Your Hypothesis

In the space below, if you need more space write in your notebook/Google docs, write down your hypothesis.

End of the Module Quiz

1. How do you form a hypothesis? Check all that apply.
   - Determine your variables
   - Prove the validity of the hypothesis
   - Generate a simple hypothesis
   - Make your prediction
   - Read and analyze your literature
   - Look for clues that answer your questions

DESIGNING YOUR EXPERIMENT

Module 5
Experimental Design Process

Designing Your Experiment

Why Design Your Experiment?

In the space below, if you need more space write in your notebook/Google docs, write down your thoughts on why it is important to design (plan out your experiment) prior to conducting the experiment.

Why Design Your Experiment?

Designing your Experiment or Making Your Experimental Plan prior to conducting your experiment is important for several reasons:

1. It ensures that you have everything that you need to conduct your experiment.
2. It ensures that you conduct your experiment safely
3. It ensures that you are organized so that you can conduct an experiment that will allow you to conduct data
4. It allows you and others to reproduce your experiments if you need to confirm or dispute experimental results.

What is Experimental Design?

Experimental Design or an Experimental Plan is the procedure that you can create and write down to conduct an experiment that tests your hypothesis. It allows you and others to know how you will conduct your experiment, what variables you will manipulate and control to measure changes in the experiment.

The Components of Your Experimental Design (Plan)

1. Materials and Equipment List

Your Material and Equipment List is a list of everything you will need to conduct your experiment for your STEM fair project. Your material and equipment list include:

- All the materials and equipment you will use to conduct your experiment
- Describe the materials and equipment in detail, i.e., the specific amount, the specific items, etc.

A Good Materials List Is Very Specific	A Bad Materials List
500 ml of de-ionized water	Water
Stopwatch with 0.1 sec accuracy	Clock
AA alkaline battery	Battery

A Sample Material List

Exercise 7: Your Material and Equipment List

In the space below, if you need more space write in your notebook/Google docs, write down the list of all the materials and equipment that you will use in your experiment. Be sure to include the amounts and units of measurements:

2. Safety Guidelines and Procedures

When conducting your experiment, you want to make sure that you are safe and those that are conducting the experiment with you are safe. The safety guidelines and procedures vary for each experiment because of the materials and equipment used in the experiment. Your safety guidelines and procedures should include the following things:

- Risks for encountering any substances while conducting your experiment
- Proper use of any equipment and/or substances
- Safely or protective equipment that should be warn
- Procedures on how to use any substances or equipment before you conduct your experiment

Exercise 8: Safety Guidelines and Procedures

In the space below, if you need more space write in your notebook/Google docs, write down the safety guidelines and procedures for your experiment.

Experimental Variables

Experimental Variables are the things that change in your experiment; they are any factor, traits or conditions that can exist in different amounts. You have 3 types of variables:

- Independent- the variable that is changed by the researcher (you). There can be only one independent variable because it allows you to better observe why the change you are seeing in your experiment is taking place.
- Dependent- the variable are what the researcher (you) focus your observations on to see how it responds to the changes made to the independent variable.
- Control-is the variable that the researcher (you) want to remain constant in the experiment.

Exercise 9: Experimental Variables

In the space below, if you need more space write in your notebook/Google docs, write down the safely guidelines and procedures for your experiment.

Step by Step Procedure

Writing your Step by Step Procedure for your experiment is like writing a step by step recipe when you are cooking. A good procedure is do detailed and complete that it allows you and/or someone else to conduct your experiment the exact same way each time your conduct the experiment. Repeating a science experiment is an important step to verify that your results are consistent and just not an accident or made up.

What should be included in your Step by Step procedure; it can include more things depending on your experiment:

- Description and Size for all experimental and control groups
- Step by Step procedure for all procedures
- Describe how you will change the independent variable and how you will measure that change
- Explain how the control variable will maintain their constant value
- How many times you intend to repeat the experiment (You should conduct any experiment at least 3 times)

Sample Step by Step Procedure

1. Number each battery so you can tell them apart.

2. Measure each battery's voltage by using the voltmeter.

3. Put the same battery into one of the devices and turn it on.

4. Let the device run for thirty minutes before measuring its voltage again. (Record the voltage in a table every time it is measured.)

5. Repeat step 4 until the battery is at 0.9 volts or until the device stops.

6. Do steps 1–5 again, three trials for each brand of battery in each experimental group.

7. For the camera flash push the flash button every 30 seconds and measure the voltage every 5 minutes.

8. For the flashlights rotate each battery brand so each one has a turn in each flashlight.

9. For the CD player repeat the same song at the same volume throughout the tests.

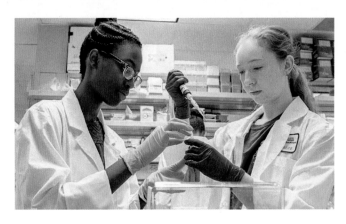

Exercise 10: Step by Step Procedure

In the space below, if you need more space write in your notebook/Google docs, write down the Step by Step Procedure for your experiment.

End of the Module Quiz

1. Match each type of variable with its definition.

Independent Variable
- The variable that is changed by you, the researcher. There is only one.

Dependent Variable
- This is the variable that you focus your observations on. It responds to changes by another variable.

Control
- This variable should remain unchanged during the experiment.

2. Your step-by-step procedures should be extremely detailed.
- True
- False

3. Your materials and equipment list should be very specific.
- True
- False

4. Why is it important to create an experimental design plan? Choose all that apply.
- It ensures you have all needed materials.
- It allows others to conduct your experiments if needed.
- It helps you stay organized as you conduct your experiment and collect data.
- It makes a note of all safety guidelines and precautions that you or anyone else need to be aware of while conducting the experiment.

CONDUCTING YOUR EXPERIMENT

Module 6
Experimental Design Process

Conducting Your Experiment

In the space below, write down what you think is needed before one conducts an experiment.

Experimental Preparation

In order to collect the best data that you can, there are a few things that you need to do prior to conducting your experiment. In the previous module, you designed your experiment. You should use that experimental plan to prepare to conduct your experiment. Before you begin to conduct your experiment, you must do the following things before you start collecting data:

- Get the materials and equipment that you need to conduct your experiment
- Determine the place where you will conduct your experiment (you must select a place that has the right experimental conditions for conducting the experiment to get great experimental results)
- Set an experimental schedule with the dates and times in which you will conduct your experiment. Make sure that you allow time in your schedule to conduct your experiment at least 3 times

Exercise 11: Experimental Plan

In the space below, if you need more space to write in a notebook or Google Docs, the answers to the questions on how you will prepare to conduct your experiment.

1. Where and when will you get the materials needed to conduct your experiment?

2. Where will you conduct your experiment? If your experiment requires you to conduct your experiment at some place other than your how or school how will you get access to that place to conduct your experiment?

3. When will you conduct your experiment? If you are working with a partner or a group be sure to coordinate your schedule so that you all can be present to conduct the experiment together?

Setting Up Your Experiment

Before you start conducting your experiment, make sure that you have the following items.

- All materials in the correct amounts to conduct your experiment
- All the required equipment to conduct your experiment
- Your Safety Guidelines, Personal Protective Equipment (PPE) and if you are working with chemicals the Material Safety Data Sheets (MSDS)
- Your written procedure
- Your Lab Notebook

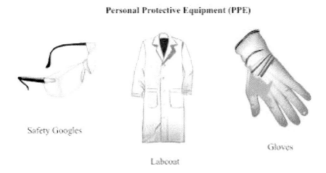

Personal Protective Equipment (PPE)

Safety Googles

Labcoat

Gloves

Exercise 12: Setting up your experiment

Before you conduct your experiment, you want to make sure that you have everything that you need to conduct your experiment in the correct place and under the correct experimental conditions (temperature, pressure, environment, etc.). In the space below, if you need more space to write use your notebook or a Google Docs, write down your experimental set up (how and where you will conduct your experiment)

Your Lab Notebook

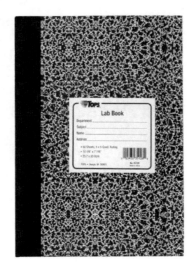

A lab notebook is what researchers use to document what happens when they are conducting an experiment. It is important to use a lab notebook because:

- You need a place to record your data and observations
- A lab notebook is a part of ensuring that your experiment is reproducible. A person should be able to read your lab notebook and conduct your experiment on their own

A good lab notebook should be:

- A bound quadrille ruled notebook
- Have everything written only in black or blue ballpoint pen
- Your project title, your name and the name of group members, contact information (just in case it gets lost and it can be returned to you) written on the outside front cover and inside front cover

How to Use A Lab Notebook

- Write Your Material and Equipment List and experimental procedure on the first few pages before you start recording data
- Write down the month, data, year that you are conducting your experiment on each page in which you are recording data in your lab notebook (the date in your notebook should be the date that you conducted your experiment)
- Record everything from your experiment in your lab notebook, data, observations, experimental conditions, etc.
- Number the pages in your notebook

What is Data?

Data is the information that you collect when conducting your experiment. Collecting data is important because it allows you to answer your experimental questions and test your hypothesis. There are two types of data:

- Qualitative data- which is data that is descriptive and subjective usually described with words.
- Quantitative data- data that can be measured and comes in numerical form usually described in tables, charts, etc.

Collecting Data and Observations

As you are conducting your experiment and collecting data, you should record everything that you do and everything that happens while you are conducting your experiment in your lab notebook, be sure to write down:

- Experimental Conditions- i.e. temperature, pressure, etc.
- Amounts of material used and the units of measurement of each amount
- Equipment used
- Anything that happened during your experiment
- Any thoughts you have about the experiment while conducting the experiment

1. What is the minimum number of times you should conduct your experiment?
   - 3
   - 2
   - 1
   - 4

2. Is it important to conduct your experiment under the correct experiment conditions?
   - Yes
   - No

3. Match each term with the correct definition.
   - Qualitative Data
     - This type of data is descriptive and subjective.
   - Quantitative Data
     - This type of data comes in numerical form.

WRITING YOUR REPORT

Module 7
Experimental Design Process

Writing Your Report

What is Data Reporting?

Data reporting is the process of collecting and presenting the data that was collected during an experiment. Data must be reported before it can be analyzed. There are 3 ways to report data:

- General Tables- a display of information in a tabular form that has columns and rows
- Graphs- visual representation of data that shows trends
- Statistical Analysis- calculate the statistics such as average, mean, standard deviation, frequency, etc.

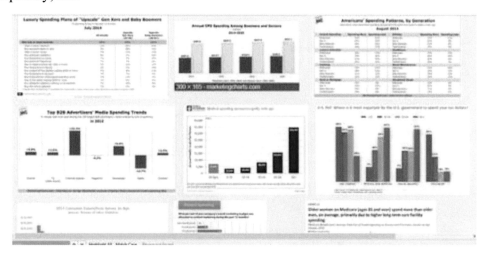

What is Data Analysis?

Data analysis is the process of reviewing and evaluating the data collected during your experiment; the data should come from what you recorded in your lab notebook as you were conducting your experiment. You should use analytical and logical reasoning as well as scientific theory to convert the data into useful information.

Why is it Important to Analyze Data?

Data analysis provides an explanation of what happened during your experiment. It determines if your experimental data supports your hypothesis. It also allows you to draw conclusions about your experiments. Here are ways to analyze and interpret your data:

- Describe and Summarize Your Data
- Identify Relationships Between Variables
- Compare Variables
- Ask if the data answers original questions and/or supports your hypothesis

Why Write a Written Report?

A written report compiles all the components of your STEM fair project, from start to finish, in one place. The purpose of the written report is to show what you did during your experiment. The information in your written report comes from the notes that you took prior to conducting your experiment (background research and step by step procedure) and your experimental data and observations that you recorded in your lab notebook.

The following items should be in your written report:

1. Title page
2. Abstract- an abstract is an abbreviated version of your final report
3. Table of Contents
4. Background Research- This is the research that you conducted prior to conducting your experiment that contains the scientific theory, key concepts and terminology connected to the STEM discipline/topic of your experiment.
5. Materials and Equipment List
6. Experimental Procedure- Your Step by Step Procedure
7. Experimental Results- Includes your data analysis which includes tables, charts, graphs and other calculations as well as any experimental observations
8. Conclusions- A summary of your results, a statement if your experimental results supported or contradicted your hypothesis, be sure to show the connections between your experiment and your theory/background research
9. Ideas for Future Research
10. Acknowledgements- Thank anyone that helped you with your experiment
11. Bibliography-A list of all the books, journal, articles, paper and websites, you read and used while conducting your background research.

What is an Abstract?

An abstract is a summary of your research paper that is usually one paragraph. It includes the major aspects of your experiment which include:

- Purpose of the Experiment
- Basic design of the Experiment
- Experimental Results

Tips for Writing Your Abstract

- Write your abstract after you've written your report
- Include your hypothesis and conclusions from your experiment
- Select key sentences and phrases from the experimental procedure
- Identify major results from your experimental results
- Make sure your abstract and paper have the same information

- Make sure you follow the abstract guidelines for the fair that you are competing in.

Sample Abstract

This study explored the pattern of video game usage and video game addiction among male college students and examined how video game addiction was related to expectations of college engagement, college grade point average (GPA), and on-campus drug and alcohol violations. Participants were 477 male, first year students at a liberal arts college. In the week before the start of classes, participants were given two surveys: one of expected college engagement, and the second of video game usage, including a measure of video game addiction. Results suggested that video game addiction is (a) negatively correlated with expected college engagement, (b) negatively correlated with college GPA, even when controlling for high school GPA, and (c) negatively correlated with drug and alcohol violations that occurred during the first year in college. Results are discussed in terms of implications for male students' engagement and success in college, and in terms of the construct validity of video game addiction.

from (www.kibin.com)

Exercise 13: Write Your Written Report

In the space write an outline for your written report. Your written report must be typed, printed and placed with your display board at the STEM fair. Be sure to use Google products (Docs and/or Sheets) or Microsoft Office (Word and/or Excel) to complete your written report.

End of the Module Quiz

1. What is the difference between data reporting and data analysis?

2. What are strategies for analyzing data? Select all that apply
   - Create Graphs
   - Summarize Your Data
   - Ask if your data answers questions and/or support your hypothesis
   - Do a Statistical Analysis

3. Why is it Important to Write an Abstract?

ORAL PRESENTATION

Module 8
Experimental Design Process

Oral Presentation

What is an Oral Presentation?

An oral presentation is a simple explanation of your project to an audience. In order to prepare for your oral presentation, you should

- Know your audience
- Determine Your Purpose for Giving Your Presentation
- Determine what multi-media format you will use with your oral presentation

Display Board or Poster for Your Oral Presentation

Depending on where you are presenting your project; you will either have to create a display board (board shaped material that is rigid and strong enough to stand on its own) or a poster (any piece of printed paper designated to be attached to a wall or vertical surface). When preparing for your STEM fair, be sure to find out if you will be required to use a display board or a poster and the acceptable dimensions of the display board or poster.

Display Board Content and Dimensions

The content for your display board will come from your written report. Here is what should go on your display board.

- Title
- Abstract
- Question
- Variables and Hypothesis
- Background Research
- Material and Equipment List
- Experimental Procedure
- Data analysis and discussion including data chart(s) & graph(s)
- Conclusions (including ideas for future research)
- Acknowledgements
- Bibliography

The standard size display board is usually 36" tall and 48" wide and are trifolds; your display board should be creative and follow the guidelines of the fair.

Preparing Your Display Board

Tips for preparing your display board:

- Don't wait until the last minute to prepare your display board
- Use quality materials and be creative
- Make a checklist on what should go on your display board
- Use pictures, graphs, tables and charts
- Make sure your title is big and easy to read from a distance
- Make sure that your font is at least 16 pt. font or larger so it can be read from a few feet away
- Check the rules for the science fair to know what is allowed on your display board and what you should have with your display board (i.e. lab notebook, written report, etc.)

Poster Content

The content for your poster will come from your written report and it should include:

- Title
- Your name, name of your collaborators (if you have any) and your institutional affiliations
- Abstract
- Background/Literature Review
- Research questions/Hypothesis
- Materials and Equipment, approach, process or methods
- Results/Conclusions
- Acknowledgments
- Contact Information

Your poster summarizes your research in a concise and attractive way. It showcases your work and facilitates a discussion with others. A poster is combination of brief text mixed with tables, graphs and pictures that is hung on a wall.

Tips for Preparing Your Poster

- Don't wait until the last minute to prepare your poster and get it printed
- Make a checklist of what should go on your poster
- Avoid clutter. Limit your poster presentation to a few main ideas.
- Keep the lettering simple. Use no more than 3 different font sizes; the largest for the poster title, second largest for the section titles and smallest for text
- Keep the colors simple
- Use pictures, graphs, tables and charts
- Check the rules for the science fair to know what is allowed on your poster

Tips for a Successful Oral Presentation

- Familiarize Yourself with the content (know your project)
- Maintain positive body language and find a good resting place for your hands
- Maintain good eye contact with the speaker
- Speak slowly and clearly

Practice Your Oral Presentation

The key to giving a successful oral presentation is practice. Write the key talking points from your experiment on notecards; practice your presentation within the time guidelines for your oral presentation. The goal is to practice until you are familiar enough with your presentation that you don't need notecards. Also ask a friend, your parents or teacher to watch your oral presentation and ask you questions so you can get used to answering questions.

End of the Module Quiz

1. How can you prepare for your oral presentation? Select all that apply.
   - Know your content.
   - Practice until you no longer need notecards.
   - Ask a friend or trusted adult to watch you as you rehearse your presentation and ask you questions.

About Ms. Tokiwa T. Smith

A native of Miami, Florida and an alumna of Florida A & M University, Ms. Tokiwa T. Smith is a Chemical Engineer, Science, Technology, Engineering and Mathematics (STEM) Educator and Social Entrepreneur. She has over 15 years' experience in education, government and philanthropy having worked for organizations such as Atlanta Public Schools, California State University East Bay, Georgia State University, Lawrence Berkeley National Laboratory and Spelman College. Ms. Smith is the Founder and Executive Director of Science, Engineering and Mathematics Link Inc., also known as SEM Link, a nonprofit organization that exposes youth to STEM and STEM Careers by connecting to the STEM community. She is also the CEO and Principal Consultant of Kemet Educational Services, a STEM educational consulting firm that focuses on ensuring that pre-college, community college and undergraduate students are prepared to pursue STEM careers.

Ms. Smith is a leader in STEM Education and Philanthropy. Currently serving on advisory boards and committees of organizations including Aerotropolis Atlanta Education Collaborative, Atlanta Public School Scientific Review Committee, Coalition for the Public Understanding of Science (COPUS) and Learn 4 Life 8th Grade Math Proficiency Change Action Network. Ms. Smith's work has been featured in several media outlets including Scientific American, US Black Engineer, Ebony Magazine, National Public Radio (NPR), Atlanta Magazine, CW Atlanta and Voyage Atlanta. Ms. Smith has been recognized by her peers in education and philanthropy by being honored by DeKalb County Public Schools with a 2019 Torch Bearer's Award for Excellence in Science Education, being selected as one of the nonprofit leaders for the 2018 Foundation Center South Boys and Men of Color Executive Director Collaboration Circle and being awarded 2018 the Black Engineer of the Year Award (BEYA) Educational Leadership Award for K-12 Promotion of Education. In 2019, she was selected as one of Atlanta Magazine's Women Making a Mark honorees. Ms. Smith is an alumna of the United Way of Greater Atlanta Volunteer Involvement Program (VIP) and a member of the National Coalition of 100 Black Women-Northwest Georgia Chapter and Junior League of Atlanta. She currently resides in Atlanta, Georgia.

Made in the USA
Monee, IL
26 May 2021